C000004019

KS1 Success

Age 5-7

Maths

Test

Practice Papers

Alan Dobbs

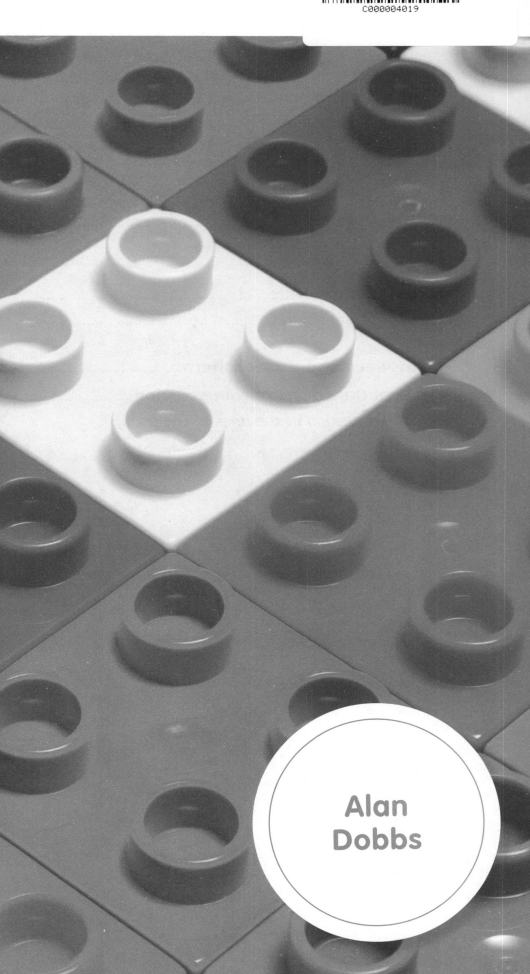

Contents

Introduction and instructions 3

Set A
Paper 1: arithmetic 5
Paper 2: reasoning 17

Set B
Paper 1: arithmetic 33
Paper 2: reasoning 45

Set C
Paper 1: arithmetic 61
Paper 2: reasoning 73

Answers and mark scheme 89
Aural questions administration 95
(pull-out section at the back of the book)

Introduction and instructions

How these tests will help your child

This book is made up of three complete sets of practice test papers. Each set contains similar test papers to those that your child will take at the end of Year 2 in maths. The tests will assess your child's knowledge, skills and understanding in the areas of study undertaken since they began Year 1. These practice test papers can be used any time throughout the year to provide practice for the Key Stage 1 tests.

The results of both sets of papers will provide a good idea of the strengths and weaknesses of your child.

Administering the tests

- Provide your child with a quiet environment where they can complete each test undisturbed.
- Provide your child with a pen or pencil and eraser. A ruler is allowed for Paper 2 but a calculator is **not** allowed in either test.
- The amount of time given for each test varies, so remind your child at the start of each one how long they have and give them access to a clock or watch.
- You should only read the instructions out to your child, not the actual questions.
- Although handwriting is not assessed, remind your child that their answers should be clear.
- Advise your child that if they are unable to do one of the questions they should go on to the next one and come back to it later, if they have time. If they finish before the end, they should go back and check their work.

Paper 1: arithmetic

- All answers are worth 1 mark, with a total number of 21 marks for each test.
- Your child will have approximately **20 minutes** to complete the test.
- The questions test your child's understanding and application of number, calculations and fractions.
- Both formal and informal methods of answering mathematical problems are acceptable.

Paper 2: reasoning

- All answers are worth 1 mark (unless stated), with a total number of 35 marks for each test.
- Your child will have approximately **35 minutes** to complete the test, including 5 minutes for the aural questions.
- The first five questions should be read out to your child. Using the aural questions administration guide on pages 95–96, read each question and remember to repeat the bold text.
- Questions will test calculation, data gathering and interpretation, shape, space and measures. They will test your child's ability to use maths in contextualised, abstract and real-life situations.

Marking the practice test papers

The answers have been provided to enable you to check how your child has performed. Fill in the marks that your child achieved for each part of the tests.

Please note: these tests are **only a guide** to the mark your child can achieve and cannot guarantee the same is achieved during the Key Stage 1 tests.

	Set A	**Set B**	**Set C**
Paper 1: arithmetic	/21	/21	/21
Paper 2: reasoning	/35	/35	/35
Total	/56	/56	/56

These scores roughly correspond with these standards: up to 15 = well below required standard; 15–24 = below required standard; 25–34 = meets required standard; over 35 = exceeds required standard.

When an area of weakness has been identified, it is useful to go over these, and similar types of questions, with your child. Sometimes your child will be familiar with the subject matter but might not understand what the question is asking. This will become apparent when talking to your child.

Shared marking and target setting

Engaging your child in the marking process will help them to develop a greater understanding of the tests and, more importantly, provide them with some ownership of their learning. They will be able to see more clearly how and why certain areas have been identified for them to target for improvement.

Top tips for your child

Don't make silly mistakes. Make sure you emphasise to your child the importance of reading the question. Easy marks can be picked up by just doing as the question asks.

Make answers clearly legible. If your child has made a mistake, encourage them to put a cross through it and write the correct answer clearly next to it. Try to encourage your child to use an eraser as little as possible.

Don't panic! These practice test papers, and indeed the end of Key Stage 1 tests, are meant to provide a guide to the standard a child has attained. They are not the be-all and end-all, as children are assessed regularly throughout the school year. Explain to your child that there is no need to worry if they cannot do a question – tell them to go on to the next question and come back to the problematic question later if they have time.

Key Stage 1

Maths

Paper 1: arithmetic

Time:

You have approximately **20 minutes** to complete this test paper.

Maximum mark	Actual mark
21	

First name	
Last name	

$$8 - 6 = \boxed{}$$

I

$$5 + 3 = \boxed{}$$

2

$$16 - 13 = \boxed{}$$

3

$$35 \div 5 = \boxed{}$$

4

$$12 + \boxed{} = 37$$

◯

5

$$10 + 11 = \boxed{}$$

◯

6

$\frac{1}{4}$ of 20 = [   ]

○

7

10 × 3 = 46 − [   ]

○

8

$$\boxed{} - 16 = 16$$

9

$$8 \times 5 = \boxed{}$$

10

$$\frac{3}{4} \text{ of } 48 = \boxed{}$$

11

$$99 = 55 + \boxed{}$$

12

$$47 = 25 + \boxed{}$$

13

$$80 \div 10 = \boxed{}$$

14

$\boxed{} - 14 = 30$

15

$25 + 16 = 41$

$41 - \boxed{} = 16$

16

$15 + 6 + 8 = $ ⬚

17

$\frac{1}{2}$ of ⬚ $= 24$

18

$30 \div 3 = 2 \times \boxed{}$

19

$18 = \dfrac{1}{2} \text{ of } \boxed{}$

20

$$30 \div \boxed{} = 2 \times 5$$

21

$$20 \times 2 = \boxed{}$$

Key Stage 1

Maths

Paper 2: reasoning

You will need to ask someone to read the instructions to you for the first five questions. These can be found on page 95. You can answer the remaining questions on your own.

Time:

You have approximately **35 minutes** to complete this test paper. This timing includes approximately 5 minutes for the aural questions.

Maximum mark	Actual mark
35	

First name	
Last name	

Practice question

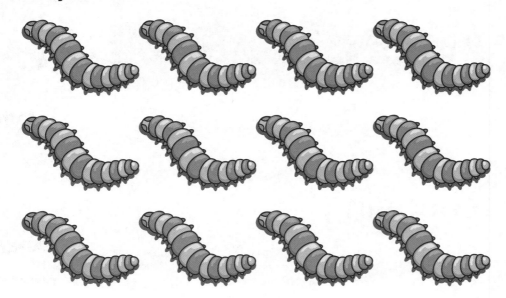

1

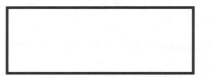

2

stickers

100 g   **3 kg**   **20 kg**   **100 kg**

4

5

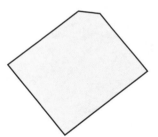

Now continue with the rest of the paper on your own.

6 Look at these signs.

$$+ \quad = \quad -$$

Write a sign in each box to make this correct.

23 ☐ 8 ☐ 15

7 This sentence is correct.

15 is more than **10** ✓

Four of these sentences are correct.
Tick (✓) them.

45 is less than **57** ☐

34 is more than **27** ☐

69 is more than **96** ☐

81 is equal to **18** ☐

28 is less than **34** ☐

59 is more than **7** ☐

8

There are **27** children.

8 children are drawing.

How many children are **not drawing**?

| |
| |
| children |

◯

9 Some of these numbers are **even**.

Put a ring around the numbers that are **even**.

13   16   29   33   76   50   27   ◯

10 Complete this table.

The first row has been done for you.

	2×5	10
		20

11 Put these numbers in order from **least** to **most**.

14   98   37   78   100   29   16

least   most

12 Circle $\frac{3}{4}$ of the bananas.

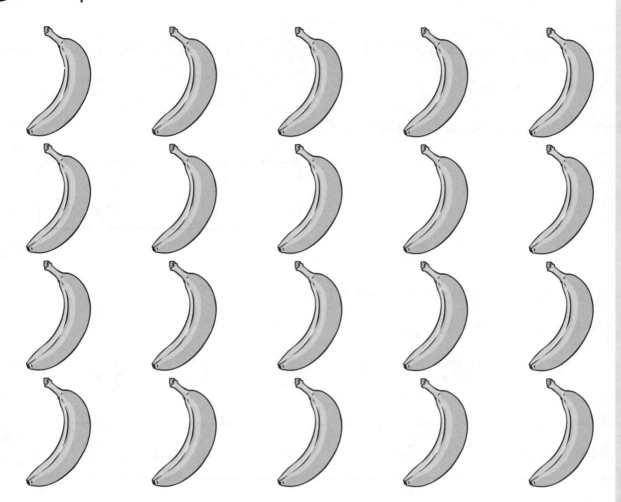

13 Shade this jug to show 650 ml.

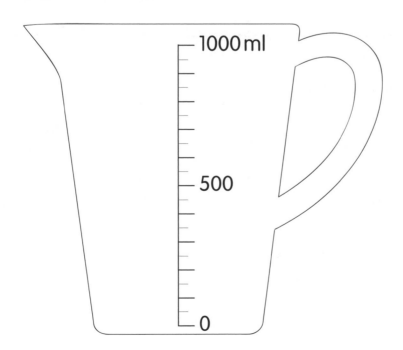

14

How many 10p coins are there in £2.55?

10p coins

15 Write each number in the correct box.

One has been done for you.

51   28   16   43   32   34

Numbers less than 30	Numbers more than 30
16	

16 Look at these cards.

Use each card once to write each of these number sentences.

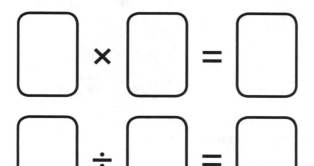

17 Tick (✓) the two shapes that are hexagons.

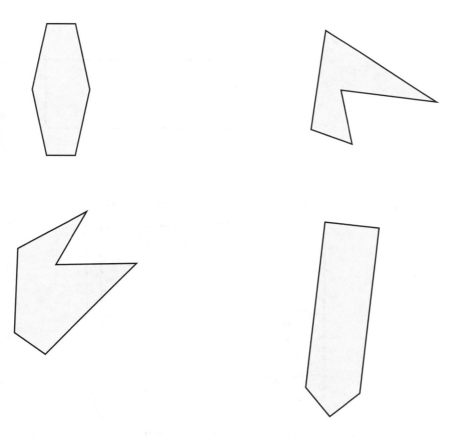

18 Pineapples cost 50p each. Oranges cost 10p each.

Adam buys **2** pineapples and **3** oranges.

How much change does he get from £1.50?

Show your working

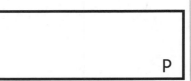

P

2 marks

19 There are **100** pieces in a kit.

14 pieces go missing.

How many pieces are left?

pieces

26

20 Class 5 have made a bar chart.

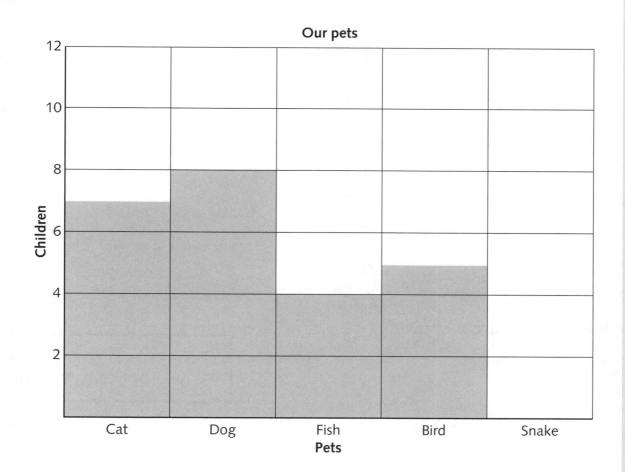

a) 3 children have a snake.

Show this on the bar chart.

b) More children have a cat than a fish.

How many more?

children

21 Complete the sequences.

a)

2	5		11		17

b)

24			15		9

22 Look at the thermometer.

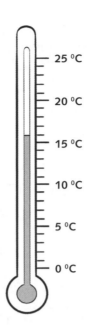

What temperature is shown?

          °C

23

 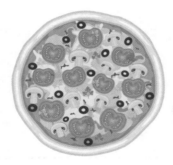 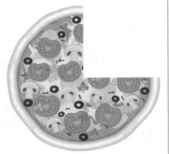

| I whole | $\frac{1}{4}$ | $\frac{3}{4}$ | $\frac{1}{2}$ |

a) Join the fractions to the correct pizza.

b) Add the total amount of pizza.

How many **whole** pizzas do you have?

whole pizzas

24 Look at the clocks.

Draw the hands on the analogue (dial) clocks so that they match the time of each digital clock.

04:30   03:00   10:45

25 Estimate the number marked by the arrow.

Write the number in the box.

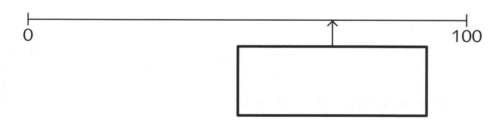

26 Look at this arrow.

You turn the arrow one clockwise quarter turn.

Tick (✓) the arrow that shows how it looks after the turn.

○

27 Match the correct name with each 3-D shape.

Cube

Cylinder

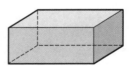

Pyramid

Cuboid

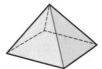

○

28 This reel has 1 m of tape.

Hamid needs to cut the tape into 20 cm pieces.

How many pieces will he have if he uses all of the tape?

pieces

○

29

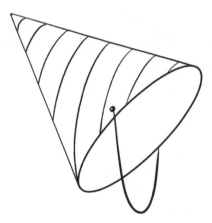

Party hats are sold in packs of 5.

Jane needs 27 hats for her party.

How many packs of hats should she buy?

packs

○

30 Mark the vertices that you can see on this cube by circling them.

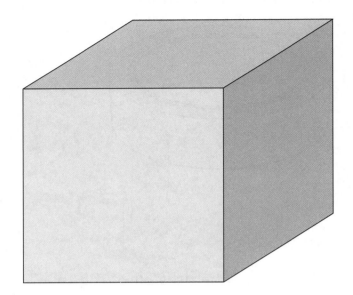

31 Circle the bee that is 6th from the left.

Key Stage 1

Maths

Paper 1: arithmetic

Time:

You have approximately **20 minutes** to complete this test paper.

Maximum mark	Actual mark
21	

First name	
Last name	

$5 + 7 = \boxed{}$

1

$34 + 12 = \boxed{}$

2

$$30 \div 3 = \boxed{}$$

○

3

$$11 + 3 + 5 = \boxed{}$$

○

4

$$\boxed{} = \frac{3}{4} \text{ of } 40$$

5

$$44 - \boxed{} = 30$$

6

$\frac{3}{4}$ of 12 = ☐

7

40 + ☐ = 100

100 = 60 + ☐

8

$$14 - \boxed{} = 7$$

9

$$10 \times 10 = \boxed{}$$

10

$\frac{1}{3}$ of 30 =

○

11

56 − 31 =

○

12

$$15 + 7 + 9 = \boxed{}$$

13

$$15 + \boxed{} = 87$$

14

Half of 20 = ☐

15

45 ÷ 5 = ☐

16

$5 \times 0 = \boxed{}$

17

$\boxed{} + 34 = 64$

18

$25 - 6 - 2 =$

19

$\dfrac{3}{4}$ of $100 =$

20

$5 + 5 + 20 = \boxed{}$

21

$67 - \boxed{} = 25$

Key Stage 1

Maths

Paper 2: reasoning

You will need to ask someone to read the instructions to you for the first five questions. These can be found on page 96. You can answer the remaining questions on your own.

Time:

You have approximately **35 minutes** to complete this test paper. This timing includes approximately 5 minutes for the aural questions.

Maximum mark	Actual mark
35	....

First name	
Last name	

Practice question

stickers

1

2

3

（empty box）

4

	faces

5

10 minutes   **30 minutes**   **2 hours**   **12 hours**

Now continue with the rest of the paper on your own.

6 Tara is standing next to her sunflower. Tara is 1 metre tall.

Tick (✓) the height that you estimate the sunflower to be.

1 metre ☐

2 metres ☐

5 metres ☐

10 metres ☐

20 metres ☐

7 Look at the grid.

One of the squares is shaded.

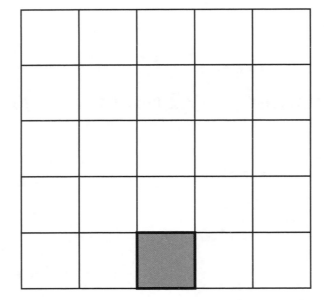

Start at the shaded square.

Move 3 squares up and 2 squares left.

Shade the square that you are now on.

8 Write the missing number.

An example has been done for you.

| 9 | double and add 5 → | 23 |

| 7 | double and add 5 → | |

9

Yasmin has emptied her money box.

She needs 56p for a set of pens.

Write 2 different ways that Yasmin could pay for the pens using her coins.

1	2

2 marks

10 Write the number that is halfway between these numbers.

| 85 | | 105 |

11 There are 50 balloons in a box.

17 balloons are green.

20 balloons are blue.

The rest of the balloons are yellow.

How many balloons are yellow?

Show your working

| | balloons |

2 marks

12 Sort the numbers into this Venn diagram.

35   16   40   15   27   36   53

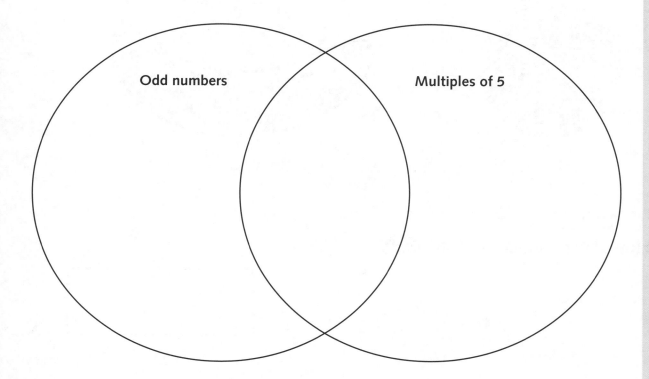

Odd numbers

Multiples of 5

13 Look at this number sequence.

	1ˢᵗ	2ⁿᵈ	3ʳᵈ	4ᵗʰ
	18	21	24	27

What would be the 7ᵗʰ number in this sequence?

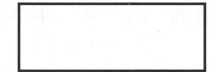

14 Ali went to the market to buy fruit.

Apples 15p   Bananas 25p

He wants to buy 2 bananas and 3 apples.

How much money will Ali need?

P

15 **Two** of these sentences describe true properties of a cylinder.

Tick (✓) the correct sentences.

Cylinders have 2 circular faces.

Cylinders have more than 4 corners.

Cylinders do not have right-angle vertices.

Cylinders have 3 triangular faces.

16 Fran made a tally of the birds visiting her garden.

Bird	Tally	Total
Robin	||||	
Finch	卌 卌 |	
Pigeon	卌	
Starling	卌 卌 ||	
Thrush	卌 |||	

Help Fran by adding the totals to her chart.

17 Look at this grid.

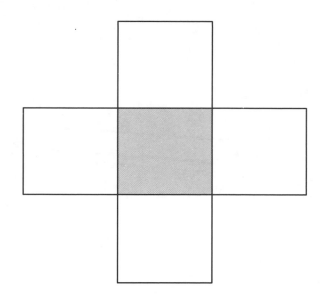

Put an X in the square above the shaded square.

Put a Y in the square below the shaded square.

18 Join the fractions with their equivalents.

One has been done for you.

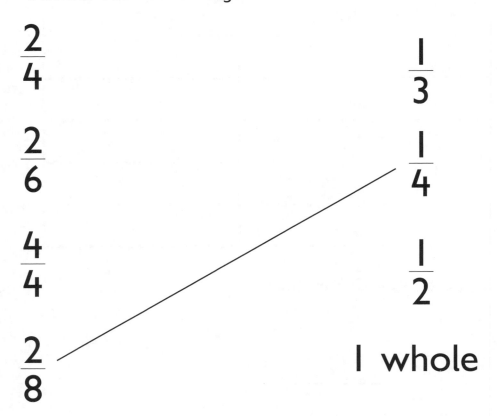

$$\frac{2}{4} \qquad \frac{1}{3}$$

$$\frac{2}{6} \qquad \frac{1}{4}$$

$$\frac{4}{4} \qquad \frac{1}{2}$$

$$\frac{2}{8} \qquad \text{I whole}$$

19 Eric wants to fill his fish tank.

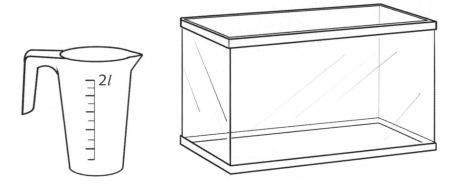

The fish tank holds 22 litres of water.

The jug holds 2 litres of water.

How many full jugs of water will Eric need?

jugs

20 Write **always**, **sometimes** or **never** in each box.

| always | sometimes | never |

| Multiples of 3 end in 1. | | |

| Multiples of 5 end in 2. | | |

| Multiples of 10 end in 0. | | |

21 Draw the reflection of this shape using the mirror line.

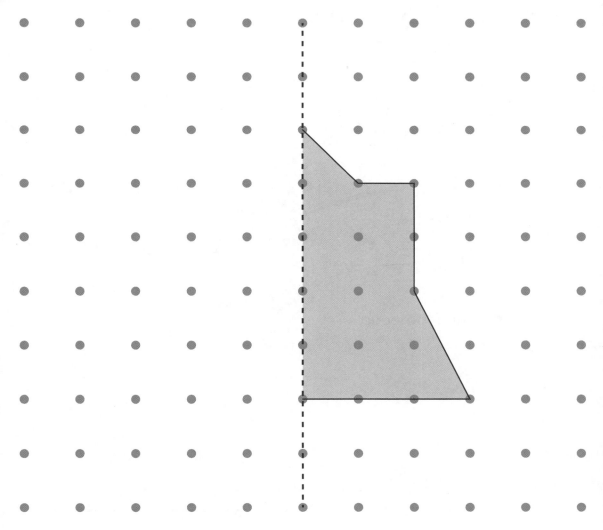

22 Add 10 to each of these numbers.

One has been done for you.

87 ──────────────▶ | 97 |

23 ──────────────▶ | |

47 ──────────────▶ | |

64 ──────────────▶ | |

23 Martina wants to make some biscuits.

For her biscuits Martina needs:

150 g flour

100 g butter

50 g dried fruit

50 g sugar

What is the total weight of Martina's ingredients?

| |
| g |

24 Gabby walks to school.

She leaves home at this time:

Her journey takes 45 minutes.

a) Show the time Gabby arrives at school on the digital clock.

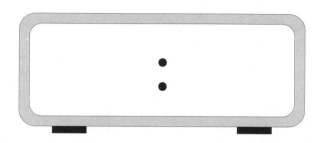

b) Draw hands on the analogue clocks to show the time that Gabby starts and finishes her journey.

Start   Finish

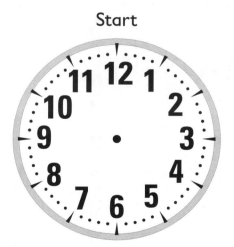   

2 marks

25 Liz's cat Smudge eats 3 small packets of food each day.

How many packets does Smudge eat in three weeks?

Show your working

packets

2 marks

26 Shema has found that 2 crayons are the same length as the top of her book.

Estimate the number of crayons that Shema would need to go around the whole book.

crayons

27 Frankie buys a burger.

He pays with a £2 coin.

Frankie gets 50p change.

How much did the burger cost?

28 Draw the next three shapes in this sequence.

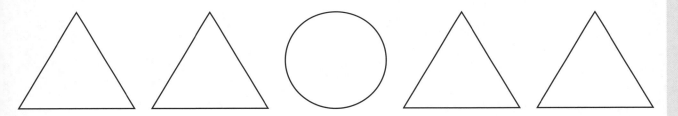

29 Use a ruler to draw a line that is 10 cm in length.

30 Write the answers in the boxes.

81 ——— take away 10 ———→ [ ]

46 ——— take away 10 ———→ [ ]

97 ——— take away 10 ———→ [ ]

Key Stage 1

Maths

Paper 1: arithmetic

Time:

You have approximately **20 minutes** to complete this test paper.

Maximum mark	Actual mark
21	

First name	
Last name	

Practice question

$3 \times 3 =$ ☐

I

$6 \div 2 =$ ☐

2

$$\boxed{} = \frac{1}{2} \text{ of } 24$$

3

$$20 - 14 = \boxed{}$$

4

| | + 12 = 38

○

5

| | × 3 = 21

○

6

$12 \div 3 =$ ◻

7

$67 - 15 =$ ◻

8

$97 - 55 = \boxed{}$

9

$\boxed{} - 9 = 19$

10

[ ] $\times\ 10 = 70$

11

[ ] $+\ 89 = 99$

12

$\boxed{} - 12 = 43$

$43 + 12 = \boxed{}$

13

$\dfrac{1}{4}$ of $20 = \boxed{}$

14

$24 \div \boxed{} = 12$

$\boxed{} \times 2 = 24$

15

$5 \times 3 = 3 + \boxed{}$

16

$\boxed{} \div 5 = 5$

17

$36 + \boxed{} = 60$

18

$\frac{1}{3}$ of 18 = [ ]

19

33 + 28 = [ ]

20

$56 - 31 =$ [ ]

21

$\dfrac{1}{3}$ of $90 =$ [ ]

Maths

Paper 2: reasoning

You will need to ask someone to read the instructions to you for the first five questions. These can be found on page 96. You can answer the remaining questions on your own.

Time:

You have approximately **35 minutes** to complete this test paper. This timing includes approximately 5 minutes for the aural questions.

Maximum mark	Actual mark
35	

First name	
Last name	

3	6	9	12	

1

| P | | P | | P |

2

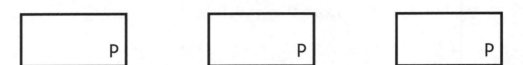

3

	children

4

	vertices

5

litres   **grams**   **centimetres**   **kilograms**

Now continue with the rest of the paper on your own.

6 Write the same number in each box to make this correct.

$$\boxed{} + \boxed{} = 30$$

7 There are 10 satsumas in each bag and 8 more loose satsumas.

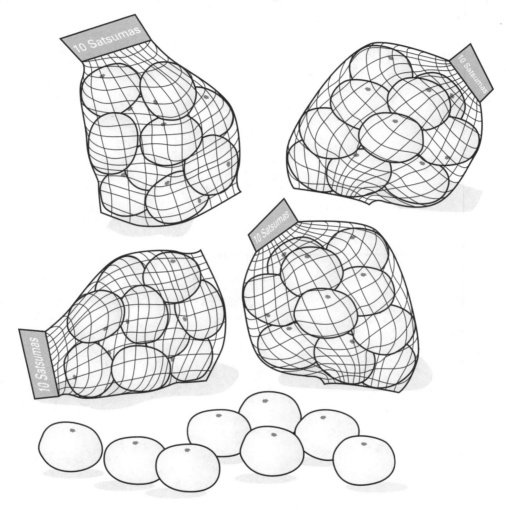

How many satsumas are there in total?

	satsumas

8 The two joined numbers add up to 18.

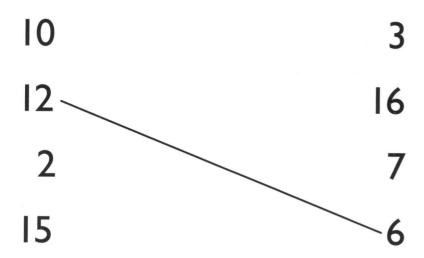

10   3

12   16

2   7

15   6

Join other numbers that have a total of 18.

9 Henry has weighed this bowl.

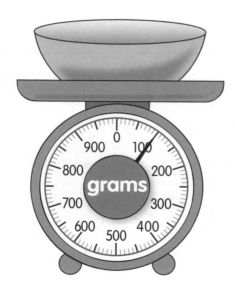

He has 5 bowls to take to his friends, so they can all have a share of ice cream.

What do Henry's bowls weigh in total?

g

10 Ben worked out the correct answer to 50 ÷ 5

His answer was 10.

Show how Ben could have worked this out.

○

11 Fill in the missing numbers.

a)

15		21		27

○

b)

55		45		35

○

12 This shape has been divided into equal parts.

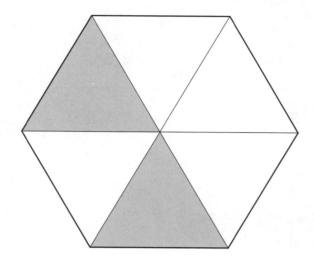

What fraction of the shape is shaded?

13 Circle the highest temperature.

20°C   22°C   18°C   21°C   12°C   ◯

14 Nina and her class have made a chart.

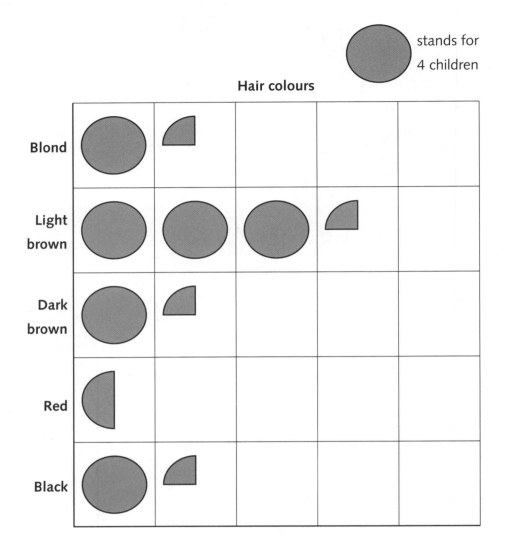

stands for
4 children

Hair colours

Blond	⬤	◔			
Light brown	⬤	⬤	⬤	◔	
Dark brown	⬤	◔			
Red	◖				
Black	⬤	◔			

a) How many children have light brown hair?

	children

b) Only one hair colour shows an **even** number of children.

Which colour is it?

15 Circle the coldest temperature.

1°C   3°C   5°C   0°C   10°C

16 Look at the number machine.

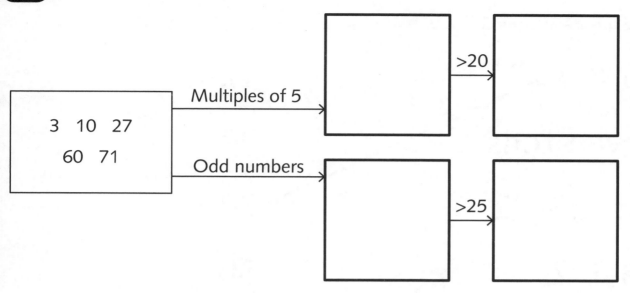

Put each number in the correct box.

17 Tara has made a list of chores.

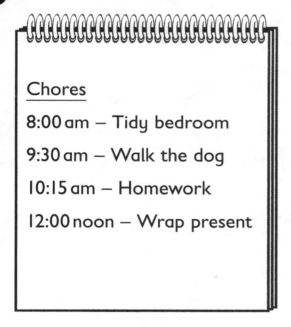

Chores

8:00 am — Tidy bedroom

9:30 am — Walk the dog

10:15 am — Homework

12:00 noon — Wrap present

How long will Tara take walking her dog?

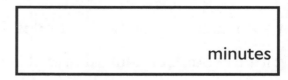

minutes

18 Draw the missing lines.

One has been done for you.

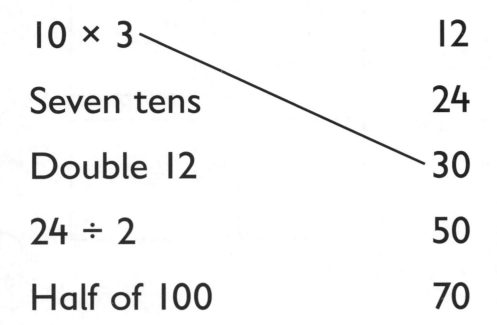

10 × 3	12
Seven tens	24
Double 12	30
24 ÷ 2	50
Half of 100	70

19 Alan is thinking of a number.

If he doubles his number and adds 7 the answer is 37.

Write the number that Alan is thinking of.

20 Starting at the flash on the far right, shade the flash that is fourth from the right.

21 Hammed has found some card shapes.
He wants to make a cuboid.

Tick (✓) the 6 shapes that he should use.

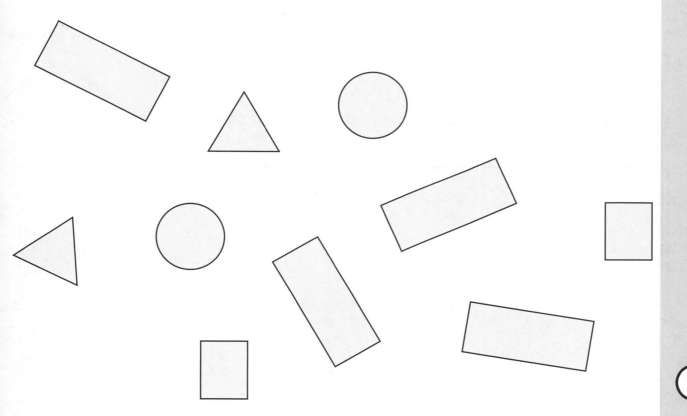

22

Alexi has emptied her money box.

She needs **87p**.

Draw the exact amount using the least amount of coins from the ones she has.

23

What is the capacity of the bottle most likely to be?

Circle the correct amount.

5 litres   **5 ml**   **2 litres**   **10 litres**   **50 ml**

24 Look at the grid.

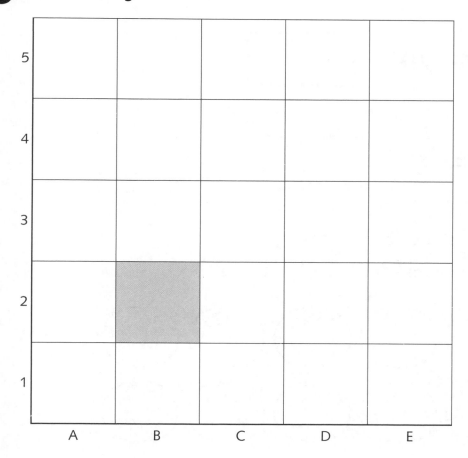

Square **B2** has been shaded.

Shade squares **C4**, **E3** and **D1**.

25

Jenny wants to buy a tablet computer that costs £90.

She saves £15 each week.

How many weeks will it take her to save enough money to buy the tablet?

weeks

26 Write the missing numbers.

a) $8 \times 10 = 94 - \boxed{}$

b) $50 \div 5 = 2 \times \boxed{}$

27 A group of children share 12 tomatoes.

Each child got 3 tomatoes.

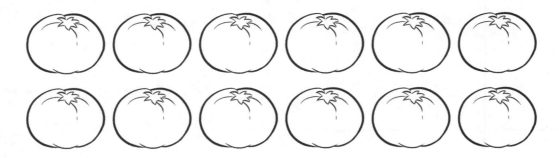

How many children shared the tomatoes?

$\boxed{}$ children

28 Shade more squares so that $\frac{3}{4}$ of the shape is shaded in total.

29 Write the numbers in the correct positions.

410   497   458

30 Class 2 have made a tally of the weather during March.

Weather in March

Weather	Tally	Total
Sunny	|||||	
Showers	||||| |||	
Cloudy	||||| |||||	
Rain	|||||	
Snow	|||	

a) Add the totals and put them in the correct column.

b) Add the missing data from the tally to the block graph.

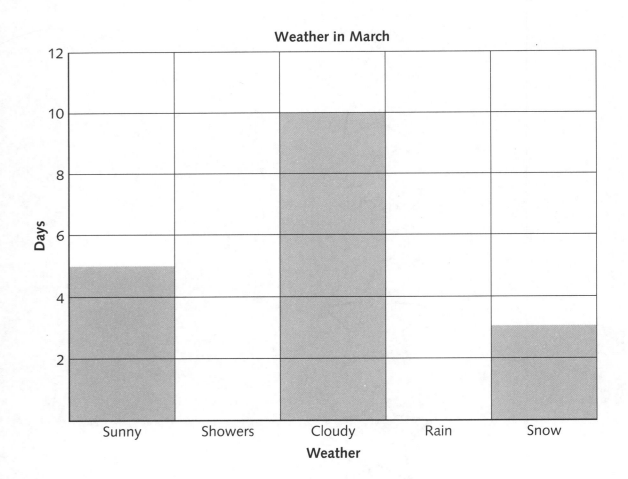

Weather in March

Answers and mark scheme

Notes to parents

Paper 1

These questions test calculating skills. An ability to work mentally will give a time advantage. Your child should be able to work with number facts to 20. (Children should know that 16 + 4 equals 20 and that 20 − 16 = 4; that 4 × 5 = 20 and that 20 ÷ 4 = 5.) Children should use this knowledge to extend their understanding of number facts to 100. Where the answer appears before the main content of the calculation, this tests your child's ability to use inverse (opposite) relationships between addition and subtraction. Knowing a fraction of a given quantity is an extension of your child's multiplication and division skills.

Paper 2

These questions test your child's ability to apply mathematics to problems and to reason, choosing an appropriate method to answer the question.

The aural section will allow your child to become accustomed to the type of questions that they should expect and will experience during the actual tests at school.

The written questions test a wide variety of skills and the ability to choose a suitable mathematical strategy to answer the problems. Some of the questions provide an extra mark for showing how the problem was solved. This is an opportunity to see how your child is using the skills that are required to become a successful mathematician. Some of the questions require a range of skills, others are more straightforward.

Set A Paper 1

Practice question   2

1.	8	(1 mark)
2.	3	(1 mark)
3.	7	(1 mark)

4.	25	(1 mark)
5.	21	(1 mark)
6.	5	(1 mark)
7.	16	(1 mark)
8.	32	(1 mark)
9.	40	(1 mark)
10.	36	(1 mark)
11.	44	(1 mark)
12.	22	(1 mark)
13.	8	(1 mark)
14.	44	(1 mark)
15.	25	(1 mark)
16.	29	(1 mark)
17.	48	(1 mark)
18.	5	(1 mark)
19.	36	(1 mark)
20.	3	(1 mark)
21.	40	(1 mark)

Set A Paper 2

Practice question   A group of 6 caterpillars circled

1.	19	(1 mark)
2.	25 stickers	(1 mark)
3.	⬭3 kg	(1 mark)
4.	45p	(1 mark)
5.		(1 mark)

6. 23 ⊟ 8 ⊟ 15   (1 mark)

7. **45** is less than **57** ✓
   34 is more than **27** ✓
   28 is less than **34** ✓
   59 is more than **7** ✓   (1 mark)

8. 19 children   (1 mark)

9. (16) (76) (50) **(1 mark)**

10.

2 × 5	10
4 × 5	20
6 × 5	30

(1 mark)

11.

14	16	29	37	78	98	100

least   most

(1 mark)

12. 15 bananas circled.

This could be 3 lots of 5, 15 as a group or any 15 bananas in any combination. **(1 mark)**

13.

(1 mark)

14. 25   10p coins **(1 mark)**

15.

Numbers less than 30	Numbers more than 30
16 28	51 43 32 34

(1 mark)

16. *Any 2 of the following multiplications and divisions:*
2 × 4 = 8   4 × 2 = 8
8 ÷ 4 = 2   8 ÷ 2 = 4 **(1 mark)**

17.

(1 mark)

18. 20p

Any method that shows the amount:

2 × 50p = £1   3 × 10p = 30p
£1 + 30p = £1.30

or

50p + 50p + 10p + 10p + 10p = £1.30

£1.50 − £1.30 = 20p

(2 marks for correct answer. Award 1 mark for using appropriate method but wrong answer given.)

19. 86 pieces **(1 mark)**

20. a) Snake column shows 3.

Note: It is important that the graph shows an understanding of each square representing 2 items and that the 3 snakes are shown as a solid lower block of 2 and a split upper block counting as 1.

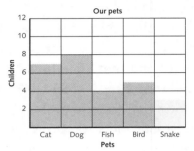

(1 mark)

b) 3 children **(1 mark)**

21. a)

2	5	8	11	14	17

(1 mark)

b)

24	21	18	15	12	9

(1 mark)

22. 16°C **(1 mark)**

23. a)

I whole $\frac{1}{4}$ $\frac{3}{4}$ $\frac{1}{2}$

(1 mark)

b) 2 whole pizzas **(1 mark)**

24.  **(1 mark)**

25. *Any number between **70** and **80** is acceptable.* **(1 mark)**

26. **(1 mark)**

27. Cube

Cylinder

Pyramid

Cuboid

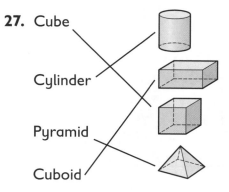

(1 mark)

28. 5 pieces (1 mark)

29. 6 packs (1 mark)

30. Vertices marked as shown:

(1 mark)

31.

(1 mark)

Set B Paper 1

Practice question 12

1. 46 (1 mark)

2. 10 (1 mark)

3. 19 (1 mark)

4. 30 (1 mark)

5. 14 (1 mark)

6. 9 (1 mark)

7. 40 + 60 = 100

100 = 60 + 40

(1 mark: both correct for 1 mark)

8. 7 (1 mark)

9. 100 (1 mark)

10. 10 (1 mark)

11. 25 (1 mark)

12. 31 (1 mark)

13. 72 (1 mark)

14. 10 (1 mark)

15. 9 (1 mark)

16. 0 (1 mark)

17. 30 (1 mark)

18. 17 (1 mark)

19. 75 (1 mark)

20. 30 (1 mark)

21. 42 (1 mark)

Set B Paper 2

Practice question 50 stickers

1. Accept 307 *or three hundred and seven*

(1 mark)

2. May (1 mark)

3. 13 (1 mark)

4. 6 faces (1 mark)

5. 2 hours (1 mark)

6. 2 metres ✓ (1 mark)

7.

(1 mark)

8. 19 (1 mark)

9. *Any 2 **different** combinations of available coins that equal 56p.*

50p + 5p + 1p

10p + 10p + 20p + 5p + 5p + 5p + 1p

or other variants.

***Do not** accept multiples of coins that are **not** available, or amounts not represented by actual coins, e.g.*

50p + 6p (*A 6p coin does not exist*)

20p + 20p + 10p + 5p + 1p (*Only 1 × 20p is available*)

(2 marks: 1 mark for each correct combination)

10. 95 (1 mark)

11. 13 balloons

Accept answers that show a possible method of calculating the answer.

20 + 17 = 37   50 − 37 = 13

50 − 30 = 20   20 − 7 = 13

(2 marks for correct answer. Award 1 mark for using appropriate method but wrong answer given.)

12.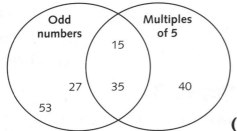

(1 mark)

13. 36 **(1 mark)**

14. 95p **(1 mark)**

15. Cylinders have 2 circular faces. ☑

Cylinders do not have right-angle vertices. ☑ **(1 mark)**

16.

Bird	Tally	Total
Robin	IIII	4
Finch	JHT JHT I	11
Pigeon	JHT	5
Starling	JHT JHT II	12
Thrush	JHT III	8

(1 mark)

17. **(1 mark)**

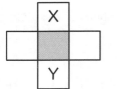

18.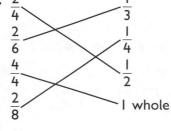

(1 mark)

19. 11 jugs **(1 mark)**

20.

Multiples of 3 end in 1.   sometimes

Multiples of 5 end in 2.   never

Multiples of 10 end in 0.   always **(1 mark)**

21.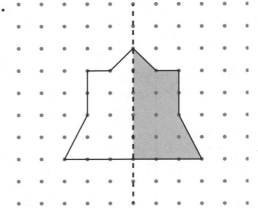

Reflective drawing shows understanding – accuracy may not be totally correct, but drawing shows concept. **(1 mark)**

22. 87 ⟶ 97

23 ⟶ 33

47 ⟶ 57

64 ⟶ 74

(1 mark)

23. 350 g **(1 mark)**

24. a) 08:15 **(1 mark)**

b) Start   Finish

(2 marks: 1 mark for each correct clock)

25. 63 packets

Any viable method shown, e.g.

7 × 3 = 21

3 × 21 = 63

3 + 3 + 3 + 3 + 3 + 3 + 3 = 21

21 × 3 = 63

(2 marks for correct answer. Award 1 mark for using appropriate method but wrong answer given.)

26. 10 crayons. *Also accept 9 or 11.* **(1 mark)**

27. £1.50 **(1 mark)**

28.  (1 mark)

29. A 10 cm line drawn correctly with a ruler.

(1 mark)

30.

81 $\xrightarrow{\text{take away 10}}$ 71

46 $\xrightarrow{\text{take away 10}}$ 36

97 $\xrightarrow{\text{take away 10}}$ 87

(1 mark)

Set C Paper 1

Practice question   9

1. 3 (1 mark)
2. 12 (1 mark)
3. 6 (1 mark)
4. 26 (1 mark)
5. 7 (1 mark)
6. 4 (1 mark)
7. 52 (1 mark)
8. 42 (1 mark)
9. 28 (1 mark)
10. 7 (1 mark)
11. 10 (1 mark)
12. $\boxed{55} - 12 = 43$
    $43 + 12 = \boxed{55}$

    (1 mark: both correct for 1 mark)
13. 5 (1 mark)
14. $24 \div \boxed{2} = 12$
    $\boxed{12} \times 2 = 24$

    (1 mark: both correct for 1 mark)
15. 12 (1 mark)
16. 25 (1 mark)
17. 24 (1 mark)
18. 6 (1 mark)
19. 61 (1 mark)
20. 25 (1 mark)
21. 30 (1 mark)

Set C Paper 2

Practice question   15

1. a) 50p
   b) 20p
   c) 2p
   in any order (1 mark)
2. 41 (1 mark)
3. 11 children (1 mark)
4. 8 vertices (1 mark)
5. (centimetres) (1 mark)
6. $15 + 15 = 30$ (1 mark)
7. 48 satsumas (1 mark)
8.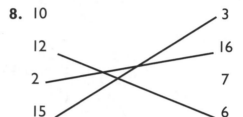

   (1 mark)
9. 500 g (1 mark)
10. *Accept:*

    10 lots of 5

    $5 + 5 + 5 + 5 + 5 + 5 + 5 + 5 + 5 + 5$

    10×5

    or any drawn answer that clearly relates to 10 groups of 5. (1 mark)
11. a)

15	18	21	24	27

(1 mark)

    b)

55	50	45	40	35

(1 mark)
12. $\frac{1}{3}$ or $\frac{2}{6}$ (1 mark)
13. (22°C) (1 mark)
14. a) 13 children (1 mark)
    b) Red (1 mark)
15. (0°C) (1 mark)

16.

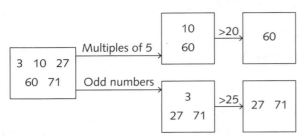

(2 marks: 1 mark for each correct row)

17. 45 minutes   **(1 mark)**

18.
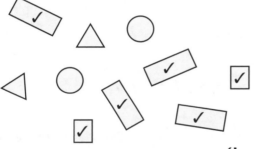

10 × 3 — 30
Seven tens — 70
Double 12 — 24
24 ÷ 2 — 12
Half of 100 — 50

(1 mark)

19. 15   **(1 mark)**

20. 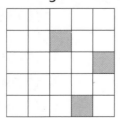   **(1 mark)**

21.

(1 mark)

22.

(1 mark)

23. (2 litres)   **(1 mark)**

24. These grid references shaded.

(1 mark)

25. 6 weeks   **(1 mark)**

26. a) 8 × 10 = 94 − [14]   **(1 mark)**

   b) 50 ÷ 5 = 2 × [5]   **(1 mark)**

27. 4 children   **(1 mark)**

28. Any 4 extra squares shaded.   **(1 mark)**

29.

523
497
481
458
410

(1 mark)

30. a)

Weather in March

Weather	Tally	Total
Sunny	|||||	5
Showers	||||| |||	8
Cloudy	||||| |||||	10
Rain	|||||	5
Snow	|||	3

(1 mark)

b) Bands shaded as shown:

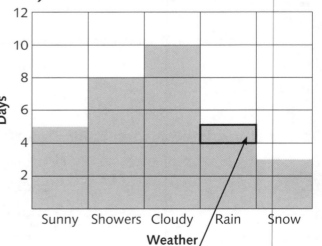

Ensure that only partial section coloured to show understanding of one block representing more than 1 item.

(1 mark)

94

Aural questions administration

The instructions below are for the first five questions in each Paper 2.

Explain to your child that the first question is a practice question. When reading the question, remember to repeat the bold text in the question. Before proceeding to question 1, ensure that your child knows where they should have written their answer and discuss with them how they worked out their answer. Allow them to change their answer to the correct one by crossing out or rubbing out, to make sure they know the way to correct errors.

Before reading out questions 1 to 5, tell your child that they should try to answer all of the questions. If they can't answer a question, they should move on to the next one and come back to that one later. Remind your child that you can't help answer these next questions and that they should try to work them out on their own. Read questions 1 to 5, allowing time for your child to write their answers. When reading the question, remember to repeat the bold text in the question. At the end of each question, allow sufficient time for your child to complete what they can.

Set A Paper 2

Read out these questions to your child. Explain that they should listen carefully and write their answers on pages 18–19.

Practice question

This is a practice question for us to do together.

Amy has 12 caterpillars. **Look at the caterpillars. Circle half of the caterpillars.**

1 Question 1. **What number is 8 less than 27?** Write your answer in the box.

2 Question 2. **There are 5 packs of stickers. Each pack has 5 stickers. How many stickers are there altogether?** Write your answer in the box.

3 Question 3. Emma's baby sister has just been born. **Circle the weight that you think Emma's baby sister is most likely to be.**

4 Question 4. **A banana costs 15p. How much money would 3 bananas cost?** Write your answer in the box.

5 Question 5. Look at the shapes. **Two of the shapes are pentagons. Tick (✓) the shapes that are pentagons.**

Set B Paper 2

Read out these questions to your child. Explain that they should listen carefully and write their answers on pages 46–47.

Practice question

This is a practice question for us to do together.

One pack has 10 stickers. How many stickers would there be in 5 packs? Write your answer in the box.

1 Question 1. **Write the number three hundred and seven.** Write your answer in the box.

2 Question 2. **February is the second month of the year. What is the fifth month?** Write your answer in the box.

3 Question 3. **What is the difference between 58 and 45?** Write your answer in the box.

4 Question 4. **A cuboid has how many faces?** Write your answer in the box.

5 Question 5. **Alba watches her favourite film. How long will the film probably last?** Circle the time that is most likely.

Set C Paper 2

Read out these questions to your child. Explain that they should listen carefully and write their answers on pages 74–75.

Practice question

This is a practice question for us to do together.

Here is a number sequence. What number would be next in this sequence? Write your answer in the box.

1 Question 1. **Hannah has 3 coins which total 72 p. Which 3 coins must Hannah have?** Write your answer in the boxes.

2 Question 2. **What is 12 less than 53?** Write your answer in the box.

3 Question 3. **There are 5 children on a bus. At the next stop 9 more children get on and 3 children get off.**

How many children are left on the bus? Write your answer in the box.

4 Question 4. **A cube has 6 faces. How many vertices does it have?** Write your answer in the box.

5 Question 5. **Max needs to know how tall he is. Which standard units will he use?** Circle the correct standard unit used to measure length and height.

KS1 Success

Age 5-7

English

Test
Practice Papers

Rachel Grant

Contents

Introduction and instructions 3

Set A

English reading Paper 1: reading prompt and answer booklet 5

English reading Paper 2: reading answer booklet 21

English grammar, punctuation and spelling Paper 1: spelling.................... 29

English grammar, punctuation and spelling Paper 2: questions 32

Set B

English reading Paper 1: reading prompt and answer booklet 44

English reading Paper 2: reading answer booklet 59

English grammar, punctuation and spelling Paper 1: spelling.................... 67

English grammar, punctuation and spelling Paper 2: questions 70

English reading booklets 81

Answers and mark scheme.. 95

Spelling test administration 103

(pull-out section at the back of the book)

Introduction and instructions

How these tests will help your child

This book is made up of two complete sets of practice test papers. Each set contains similar test papers to those that your child will take at the end of Year 2 in English reading and English grammar, punctuation and spelling. The tests will assess your child's knowledge, skills and understanding in the areas of study undertaken since they began Year 1. These practice test papers can be used any time throughout the year to provide practice for the Key Stage 1 tests.

The results of both sets of papers will provide a good idea of the strengths and weaknesses of your child.

Administering the tests

- Provide your child with a quiet environment where they can complete each test undisturbed.
- Provide your child with a pen or pencil, ruler and eraser.
- The amount of time given for each test varies, so remind your child at the start of each one how long they have and give them access to a clock or watch.
- You should only read the instructions out to your child, not the actual questions.
- Although handwriting is not assessed, remind your child that their answers should be clear.
- Advise your child that if they are unable to do one of the questions they should to go on to the next one and come back to it later, if they have time. If they finish before the end, they should go back and check their work.

English reading

Paper 1: reading prompt and answer booklet

- Each test is made up of an answer booklet containing two different reading prompts and questions.
- All answers are worth 1 mark, with a total number of 20 marks for each test.
- Your child will have approximately **30 minutes** to read the prompts and answer the questions.
- Read the list of useful words and discuss their meanings with your child.
- Read the practice questions out loud to your child and allow them time to write down their own answer.
- Your child should read and answer the other questions by themselves.

Paper 2: reading answer booklet

- Each test is made up of two different texts and an answer booklet.
- All answers are worth 1 mark (unless otherwise stated), with a total number of 20 marks for each test.
- Your child will have approximately **40 minutes** to read the texts in the reading booklet and answer the questions in the answer booklet.

English grammar, punctuation and spelling

Paper 1: spelling

- Contains 20 spellings, with each spelling worth 1 mark.
- Your child will have approximately **15 minutes** to complete the test paper.
- Using the spelling administration guide on pages 103–104, read each spelling and allow your child time to fill it in on their spelling paper.

Paper 2: questions

- All answers are worth 1 mark (unless otherwise stated), with a total number of 20 marks for each test.
- Your child will have approximately **20 minutes** to complete the test paper.
- Read the practice questions out loud to your child and allow them time to write down their own answer.
- Some questions are multiple choice and may require a tick in the box next to the answer. Some require a word or phrase to be underlined or circled while others have a line or box for the answer. Some questions ask for missing punctuation marks to be inserted.

Marking the practice test papers

The answers and mark scheme have been provided to enable you to check how your child has performed. Fill in the marks that your child achieved for each part of the tests.

Please note: these tests are **only a guide** to the mark your child can achieve and cannot guarantee the same is achieved during the Key Stage 1 tests.

English reading

	Set A	Set B
Paper 1: reading prompt and answer booklet	/20	/20
Paper 2: reading answer booklet	/20	/20
Total	/40	/40

These scores roughly correspond with these standards: up to 10 = well below required standard; 11–20 = below required standard; 21–30 = meets required standard; over 31 = exceeds required standard.

English grammar, punctuation and spelling

	Set A	Set B
Paper 1: spelling	/20	/20
Paper 2: questions	/20	/20
Total	/40	/40

These scores roughly correspond with these levels: up to 10 = well below required level; 11–20 = below required level; 21–30 = meets required level; over 31 = exceeds required level.

When an area of weakness has been identified, it is useful to go over these, and similar types of questions, with your child. Sometimes your child will be familiar with the subject matter but might not understand what the question is asking. This will become apparent when talking to your child.

Shared marking and target setting

Engaging your child in the marking process will help them to develop a greater understanding of the tests and, more importantly, provide them with some ownership of their learning. They will be able to see more clearly how and why certain areas have been identified for them to target for improvement.

Top tips for your child

Don't make silly mistakes. Make sure you emphasise to your child the importance of reading the question. Easy marks can be picked up by just doing as the question asks.

Make answers clearly legible. If your child has made a mistake, encourage them to put a cross through it and write the correct answer clearly next to it. Try to encourage your child to use an eraser as little as possible.

Don't panic! These practice test papers, and indeed the Key Stage 1 tests, are meant to provide a guide to the standard a child has attained. They are not the be-all and end-all, as children are assessed regularly throughout the school year. Explain to your child that there is no need to worry if they cannot do a question – tell them to go on to the next question and come back to the problematic question later if they have time.

Key Stage 1

English reading

Paper 1: reading prompt and answer booklet

Time:

You have approximately **30 minutes** to complete this test paper.

Maximum mark	Actual mark
20	

First name	
Last name	

Useful words

lighthouse

souvenir

The Story of Grace Darling

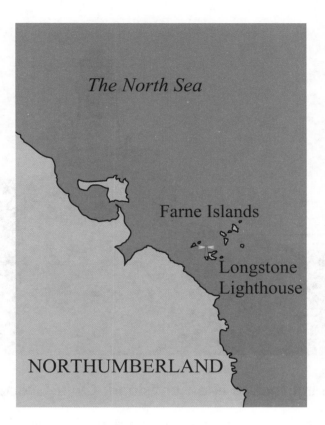

The North Sea

Farne Islands

Longstone
Lighthouse

NORTHUMBERLAND

Grace Darling was a lighthouse keeper's daughter. She lived in Longstone
Lighthouse, which is in Northumberland in north-east England.

Practice questions

a Who was Grace Darling?

She was the lighthouse keepers daughter. ✓

b **Find** and **copy** the name of the lighthouse where Grace Darling lived.

Longstone Lighthouse. ✓

The lighthouse is on an island. Only Grace's family lived on the island.

Grace's father had an important job. He kept the light burning in the lighthouse.

The light warned ships about dangerous rocks in the sea.

---

1 What did Grace's father do?

He kept the light burning in the lighthouse ✓

2 Which word in the text describes what the rocks were like?

Tick **one**.

lonely ☐	important ☐
dangerous ☑	burning ☐

On 7 September 1838 there was a big storm.

A ship was passing Little Farne Island. It had **63** people on board.

Suddenly, the engine stopped. The crew could not use the sails because of the storm.

The ship drifted and hit the rocks.

---

3 How many people were on board the ship?

Tick **one**.

7 ☐   63 ☑

18 ☐   38 ☐

4 Give **two** reasons why the ship drifted and hit the rocks.

1. _the storm was so strong._

2. _the engine stopped._ ✓

Grace Darling saw the accident. She told her father about it. They did not think anyone on the ship would be alive.

When the sun rose, they saw people clinging to the rocks. Some people were still alive!

---

5 What did Grace do when she saw the accident?

She told her dad. ✓

○

6 What did Grace and her father see when the sun rose? ✗

into They saw a boat crashing into rocks. ○

Grace and her father took their tiny rowing boat. They rowed through the stormy sea, in winds and rain. They found nine people on the rocks. Grace kept control of the boat while her father went to help.

One by one, they helped the people into the boat. They rowed them to safety.

---

7 **Find** and **copy one** word that tells you what the sea was like.

Stormy ~~Sea~~ ✓

8 Number the pictures 1 to 4 to show the order they happened in the story.

The first one has been done for you.

When people heard Grace Darling's story, she became a national hero. Even Queen Victoria heard the story. Grace was given a silver medal.

Poems and songs were written about Grace. Souvenirs such as postcards and mugs with pictures of Grace on them were made.

---

9 Why did Grace become a national hero?

Tick **one**.

People wrote songs about Grace. ☐

People thought Grace was very brave. ☑

Grace's picture was on postcards. ☐

Queen Victoria heard Grace's story. ☐

○

Useful words

direction

thousands

Why Owls have Big Eyes

Owl and Pigeon were friends. They liked to meet at dawn and watch the sunrise. They liked to talk about many things.

One morning, Owl said to Pigeon, "There are more owls than pigeons."

"Nonsense!" said Pigeon. "There are lots more pigeons than owls!"

Practice questions

c *They liked to meet at dawn and watch the sunrise.*

What does *at dawn* mean in this sentence?

Tick **one**.

get together ☐   at home ☐

early in the morning ☐   in the woods ☐

d What did Owl say to Pigeon?

"There are _____"

"There's only one way to find out," hooted Owl. "We will count them!"

"All right," cooed Pigeon. "Where shall we do it? We need a big place."

Owl thought for a minute and then he said, "The Big Wood has lots of trees."

"Fine," said Pigeon. "You tell the owls and I'll tell the pigeons. One week from today we will count them."

They both flew off to tell everyone.

10 Why was the Big Wood a good place to count the owls and pigeons?

_____ ◯

11 When did they decide they would count the owls and pigeons?

Tick **one**.

in two weeks	☐	today ☐
later this week	☐	in one week ☐ ◯

One week later, the owls arrived at the Big Wood, just as the sun rose.

They flew down from every direction until the trees were full.

They laughed and said, "Too-woo, wah, wah! The pigeons are still asleep!"

They were sure there could not be as many pigeons.

---

12 How do you know there were a lot of owls in the Big Wood?

_____

⭘

13 *They were sure there could not be as many pigeons.*

What does the word *sure* mean in this sentence?

Tick **one**.

certain ☐   safe ☐

trust ☐   truthful ☐   ⭘

After a while they saw huge grey clouds moving towards them. The clouds were pigeon wings.

Thousands of pigeons flew down to the Big Wood and as they landed, the owls moved closer together. Branches broke when too many pigeons tried to land at once.

The owls' eyes grew wider and wider as they tried to see all the pigeons. They stared and moved their heads from side to side to watch the pigeons.

14 **Find** and **copy two** words that tell you what the clouds really were.

_____ ◯

15 Why did the owls move closer together?

Tick **one**.

To laugh at the pigeons. ☐   To make room for the pigeons. ☐

To watch the pigeons. ☐   To welcome the pigeons. ☐ ◯

The pigeons kept coming. The owls could not believe there were so many pigeons!

The owls started to feel nervous and hooted, "Tooooo-woooo! Let's get out of here!"

One by one the owls flew away, up between the branches.

---

16 Why did the owls decide to fly away?

Tick **two**.

It was getting late.	☐	They felt cold.	☐
They felt nervous.	☐	There was no room for them.	☐

○

17 Number the sentences below from 1 to 4 to show the order they happen in the story.

The first one has been done for you.

The owls flew off.	☐	The owls became worried.	☐
More pigeons arrived.	1	The owls were surprised.	☐

○

Owl and Pigeon never did count how many owls and pigeons there were.

Now the owls always fly at night so that they will not meet pigeons.

Their eyes are big and round and they stare because they are looking out for pigeons.

---

18 Do you think there were more pigeons than owls?

Tick **one**.

Yes ☐   No ☐

Give one reason for your answer.

_____ ◯

19 Give **two** things that owls do now because of the pigeons.

1. _____

2. _____ ◯

20 Think about all you have read.

How do you think the pigeons felt after the owls flew away, and why?

_____

_____

Key Stage 1

English reading

Paper 2: reading answer booklet

Time:

You have approximately **40 minutes** to read the texts
in the reading booklet (pages 82–87) and answer the
questions in the answer booklet.

Maximum mark	Actual mark
20	

First name	
Last name	

(page 82)

1 Why did Anansi want to taste the food that his neighbours cooked?

Because he was greedy ①

(page 82)

2 *they were stubby, but strong.*

What does the word *stubby* mean?

Tick **one**.

black and hairy ☐

short and thick ☐

long and thin ☒

tired and weak ☑ ⓪

(page 82)

3 Why didn't Anansi wait at Rabbit's house?

Because Anansi wanted some Soup. ⓪

(page 83)

4 What was Fox cooking for his family?

He was cooking beans and rice ①

22

(page 83)

5 At Monkey's house, why did Anansi tie his leg to the oven door?

<u>He ~~tied~~ tied it to the oven ~~because~~</u>
<u>then Anansi ~~could not new~~ when to go there</u>

(page 84)

6 **a)** Draw lines to match these characters to the foods they were cooking.

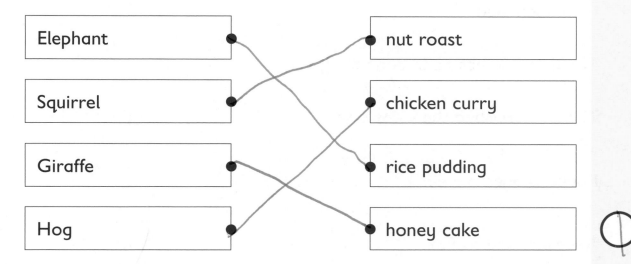

Elephant	nut roast
Squirrel	chicken curry
Giraffe	rice pudding
Hog	honey cake

b) What did each of the animals promise to do when dinner was ready?

<u>So Anansi could ~~coe~~ come to the house</u>

(page 84)

7 **Find** and **copy** the **three** words that show that Anansi was looking forward to having eight dinners.

~~Great~~, Great delicious

Promisd

(pages 82–85)

8 Number the sentences below from 1 to 4 to show the order in which they happened.

The first one has been done for you.

The river washed the webs away.   4

Rabbit tugged a web.   2

Anansi talked to Fox.   1

Anansi's legs started to grow long and skinny.   3

24

(page 86)

9 Water is needed for...

Tick **two**.

washing our clothes. ☐

burning wood. ☐

plants to grow. ☐

drying our clothes. ☐

○

(page 86)

10 Look at the section headed: **The Water Cycle**

Find and **copy** the word that means the same as *small drops*.

_____

○

(page 86)

11 What does the word *cycle* mean in *the water cycle?*

_____

_____

○

25

12 Copy the two labels into the correct boxes on either side of the diagram.

| water changes from liquid into gas | water falls as rain, hail, snow or sleet |

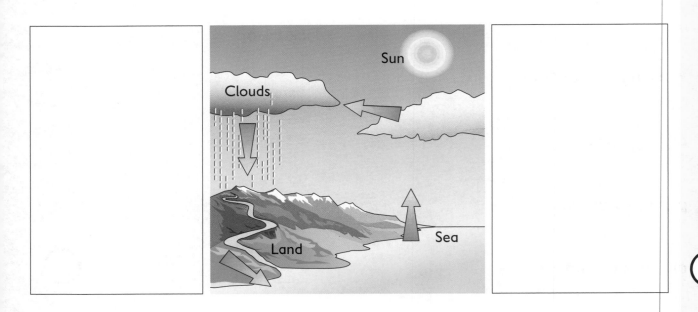

(page 86)

13 What happens to water when the sun heats it up?

_____

_____

14 Look at the section: **FACT FILE**

Find and **copy** how much water you have in your:

1. blood _____

2. skin _____ ◯

15 Look at the section: **What is Polluted Water?**

Give **three** things that can cause water pollution.

1. _____

2. _____

3. _____ ◯

(page 87)

16 **Find** and **copy two** words that mean the same as *rubbish*.

I. _____

2. _____

◯

(page 87)

17 Why is water precious?

Give **one** reason.

_____

◯

(page 87)

18 Put ticks in the table to show which sentences are **true** and which are **false**.

One has been done for you.

The information says that...	True	False
humans cause all polluted water.		✓
we should try to save water.		
90% of the world's water is not fresh.		
you should pick up your litter, but only when you are at a beach, lake or river.		
a watering can uses less water than a hosepipe.		

◯

2 marks

END OF TEST

28

Key Stage 1

English grammar, punctuation and spelling

Paper 1: spelling

You will need to ask someone to read the instructions and sentences to you. These can be found on page 103.

Time:

You have approximately **15 minutes** to complete this test paper.

Maximum mark	Actual mark
20	

First name	
Last name	

Spelling

Practice question

The children _____ in the park.

---

1. The ice is too thin to _____ on.

2. Our cat has long _____.

3. An elephant's nose is called a _____.

4. Sit in the chair _____ the door.

5. Sally found a magpie _____.

6. Please give me your _____ number.

7. We all enjoyed the cricket _____.

8. Dad _____ Milly's pram.

9. We are going swimming on _____.

10. We build a _____ on the beach.

11. Hannah has _____ to do today.

12. Jack takes the _____ to the Post Office.

13. You wear a watch on your _____.

14. The hat is _____ with yellow flowers.

15. I do not want to _____ with you.

16. Please pass me the pink _____.

17. This fish is very _____.

18. Plastic is a _____ material.

19. Our school has a _____ in September.

20. Alex gave me an _____ to his party.

END OF SPELLING TEST

Key Stage 1

English grammar, punctuation and spelling

Paper 2: questions

Time:

You have approximately **20 minutes** to complete this test paper.

Maximum mark	Actual mark
20	

First name	
Last name	

Practice questions

a Write one word on the line to complete the sentence in the present tense.

Sally is _____ a book.

b Tick the correct punctuation mark to complete the sentence.

Where do you live

Tick **one**.

full stop ☐

question mark ☐

exclamation mark ☐

comma ☐

1 Circle **three** words that should have a capital letter in the sentences below.

it is a hot day. sally and tom are eating ice cream.

○

2 Tick the correct word to complete the sentence below.

Do you prefer orange juice _____ apple juice?

and [ ]

or [ ]

but [ ]

so [ ]

○

3 Tick the phrase that uses the correct punctuation to complete the sentence below.

Where are you going _____

Tick **one**.

on holiday this year! ☐

on holiday this year. ☐

on holiday this year ☐

on holiday this year? ☐

○

4 Tick the punctuation mark that should complete each sentence.

Sentence	Full stop	Question mark
You can have tea at our house		
Are you coming to play football		
Where are my shoes		
It is lunchtime now		

○

5 What type of word is underlined in the sentence below?

Meera carefully picked up the fluffy <u>hamster</u>.

Tick **one**.

a verb ☐

a noun ☐

an adjective ☐

an adverb ☐

6 Write the missing punctuation mark to complete the sentence below.

What a wonderful surprise

7 Read the sentences below.

Last week, we went to the seaside.

We took our buckets and spades and we packed a picnic.

We played on the sandy beach and in the rock pools.

Then we paddled in the warm sea.

Tick the word that best describes these sentences.

Tick **one**.

questions ☐

commands ☐

exclamations ☐

statements ☐

8 Circle the **four** nouns in the sentence below.

Green plants need air, water and sunlight to grow.

9 Which word correctly completes the sentence?

Brett felt _____ when he lost his favourite toy.

Tick **one**.

happy ☐

inhappy ☐

unhappy ☐

imhappy ☐

10 Underline the **two** adjectives in the sentence below.

In the garden, a cheeky robin hopped onto a branch and he sang a sweet song.

○

11 Write **ful** or **less** to complete the word in each sentence.

Did you see that colour_____ rainbow in the sky?

The fear_____ explorer went bravely into the jungle.

○

12 What type of word is underlined in the sentence below?

The fat ginger cat gazed <u>hungrily</u> at the bright blue fish darting in the water.

an adverb ☐

a noun ☐

an adjective ☐

a verb ☐

13 Write **three** commas in the correct places in the sentence below.

You must bring a coat gloves a hat boots and a packed lunch to school on Friday.

14 The verb in the box is in the present tense.

Write the **past tense** of the verb in the space.

$$\boxed{\text{runs}}$$
↓

Aisha's mum _____ very fast in the egg and spoon race.

◯

15 Why does the underlined word have an apostrophe?

Put your books on the teacher's desk.

_____

_____

◯

16 Write **was** or **were** to complete the sentence below correctly.

The birds _____ flying high up in the sky.

◯

17 Yusuf and Brandon are at the zoo. They are looking at the parrots. Brandon wants to give the parrots some of his sandwich. Yusuf wants to stop him.

Write the words that Yusuf says to Brandon in the speech bubble.

Remember to use the correct punctuation.

2 marks

18 The verbs that are underlined are in the present tense.

Write these verbs in the **past tense**.

One has been done for you.

Hari <u>bakes</u> a cake for my birthday.

Hari <u>baked</u> a cake for my birthday.

He <u>uses</u> flour, butter, eggs and sugar.

He _____ flour, butter, eggs and sugar.

I <u>choose</u> the chocolate icing and cherries to go on top.

I _____ the chocolate icing and cherries

to go on top.

2 marks

END OF TEST

Key Stage 1

English reading

Paper 1: reading prompt and answer booklet

Time:

You have approximately **30 minutes** to complete this test paper.

Maximum mark	Actual mark
20	

First name	
Last name	

Key Stage 1

SET
B

English
grammar,
punctuation
and spelling

PAPER 1

English grammar, punctuation and spelling

Paper 1: spelling

You will need to ask someone to read the instructions and sentences to you. These can be found on pages 103–104.

Time:

You have approximately **15 minutes** to complete this test paper.

Maximum mark	Actual mark
20	....

First name	
Last name	

Spelling

Practice question

The birds _____ into the sky.

_____

1. Gina shows great _____ on the piano.

2. The postman brought a big _____.

3. The eggs _____ under the hen.

4. Our new _____ loves to play with us.

5. The children had an amazing time at the _____.

6. The sky was _____ when we left home.

7. Tom _____ TV after he has done his homework.

8. Dad is _____ vegetables for dinner.

9. Gran is coming to tea on _____.

10. Jane was too _____ to join in our game.

11. He thinks reading is _____ than writing.   ◯

12. Eating fresh fruit every day is good for your _____.   ◯

13. The hot air balloon rose _____ the rooftops.   ◯

14. She loves all vegetables, especially _____.   ◯

15. Mum _____ to answer the door.   ◯

16. Jane _____ at school today.   ◯

17. Uncle Toby took his _____ to the park.   ◯

18. The river runs through a deep _____.   ◯

19. Yesterday I saw a _____ with seven spots.   ◯

20. I enjoy reading stories with lots of _____.   ◯

END OF SPELLING TEST

SET
B

English
grammar,
punctuation
and spelling

PAPER 2

Key Stage 1

English grammar, punctuation and spelling

Paper 2: questions

Time:

You have approximately **20 minutes** to complete this test paper.

Maximum mark	Actual mark
20	

First name	
Last name	

Practice questions

a Tick the word that completes the sentence.

Gerbils are usually _____ than hamsters.

smallest ☐

small ☐

little ☐

smaller ☐

b Write the words **do not** as one word, using an apostrophe.

Please _____ pick the flowers.

1 Underline the noun phrase in the sentence.

Do you like my new green hat?

2 Circle **three** words that must have capital letters in the sentence.

My birthday is in april and jane's birthday is in may.

3 Tick the best option to complete the sentence.

Will you put the books away, please

full stop ☐

exclamation mark ☐

question mark ☐

comma ☐

4 Write the missing punctuation mark to complete the sentence below.

Our school trip is to Leeds Castle

○

5 What type of word is <u>hurt</u> in the sentence below?

Shelley slipped on the ice and hurt her leg.

Tick **one**.

a noun ☐

an adjective ☐

an adverb ☐

a verb ☐

○

6 Which sentence has the correct punctuation?

Tick **one**.

What a great idea ☐

What a great idea! ☐

What a great idea? ☐

What a great idea. ☐

7 Circle the adverb in the sentence below.

We all cheered loudly when my sister won the race.

8 Tick the word that completes the sentence.

They _____ at a lovely hotel.

Tick **one**.

stayed ☐

stays ☐

staying ☐

to stay ☐

9 Tick to show whether each sentence is in the **present tense** or the **past tense.**

	Present tense	Past tense
I was talking to Mary.		
Jack is brushing his teeth.		

10 Read the sentences below.

Oh no! The dog has chewed my trainers!

Tick the word that best describes the sentences.

statements ☐

exclamations ☐

commands ☐

questions ☐

11 Tick the sentence that is a **statement**.

Tick **one**.

How exciting! We're going on holiday! ☐

Have you got the tickets? ☐

Carry your suitcase to the car. ☐

I have locked the front door. ☐

12 Write **two** commas where they should go in the sentence below.

Spring summer autumn and winter are the four seasons of the year.

○

13 Tick the sentence that is written correctly in the past tense.

Tick **one**.

We are going to the shop and buying some ice cream. ☐

We went to the shop and bought some ice cream. ☐

We go to the shop and bought some ice cream. ☐

We went to the shop and buy some ice cream. ☐ ○

14 Circle **two** adjectives in the sentence below.

When we woke up the next morning, the soft, white snow was falling silently from the sky.

15 Write the words in the boxes as one word, using an apostrophe.

One has been done for you.

I am

I'm building a sandcastle.

can not

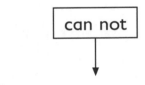

He _____ go swimming today.

he will

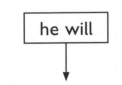

I hope _____ be better tomorrow.

2 marks

16 Tick **two** sentences that are correct.

They is closing the paddling pool. ☐

The park is closing soon. ☐

Peter is closing the door quietly. ☐

I are closing my eyes and counting to ten. ☐

○

17 Draw lines to match the sentences with their correct type.

When we got home it was time for bed. ●	● question
Pass the salt. ●	● exclamation
When is it lunchtime? ●	● statement
Thank you! I love it! ●	● command

○

18 The verbs in the boxes are in the **present tense.**

Write the verbs in the correct tense in the sentences below.

One has been done for you.

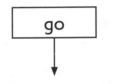

The children were shouting and the dog ___was barking___ at the door.

Mum _____ to work but she couldn't think because of the noise.

She _____ downstairs to find out what was wrong.

2 marks

END OF TEST

Answers and mark scheme

Set A English reading

Paper 1: reading prompt and answer booklet

The Story of Grace Darling

a a lighthouse keeper's daughter

b Longstone Lighthouse

1 He kept the light burning in the lighthouse. **(1 mark)**

2 lonely ☐ important ☐

dangerous ☑ burning ☐ **(1 mark)**

3 7 ☐ 63 ☑

18 ☐ 38 ☐ **(1 mark)**

4 Any two from: the engine stopped; the crew could not use the sails; because of the storm.

(1 mark: two correct for 1 mark)

5 Grace/She told her father about it. **(1 mark)**

6 Any answers that refer to people clinging to the rocks/people still alive. **(1 mark)**

7 stormy **(1 mark)**

8

2	1
3	4

(1 mark: award 1 mark for the pictures correctly numbered)

9 People wrote songs about Grace. ☐

People thought Grace was very brave. ☑

Grace's picture was on postcards. ☐

Queen Victoria heard Grace's story. ☐

(1 mark)

Why Owls have Big Eyes

c get together ☐

at home ☐

early in the morning ☑

in the woods ☐

d more owls than pigeons

10 Any one from: (because) the Big Wood/it had lots of trees; (because) they needed a big place. **(1 mark)**

11 in two weeks ☐ today ☐

later this week ☐ in one week ☑ **(1 mark)**

12 Because the trees were full. **(1 mark)**

13 certain ☑ safe ☐

trust ☐ truthful ☐ **(1 mark)**

14 pigeon wings **(1 mark)**

15 To laugh at the pigeons. ☐

To make room for the pigeons. ☑

To watch the pigeons. ☐

To welcome the pigeons. ☐ **(1 mark)**

16 It was getting late. ☐ They felt cold. ☐

They felt nervous. ☑ There was no room for them. ☑

(1 mark: award 1 mark for two correct answers ticked)

17 The owls flew off. ☐4 The owls became worried. ☐3

More pigeons arrived. ☐1 The owls were surprised. ☐2

(1 mark: award 1 mark for correct order)

18 Any of the following answers are acceptable.

Yes, because:

- the owls were surprised to see so many pigeons.
- the owls felt nervous because there were so many pigeons.
- the owls flew away because there were more pigeons than owls.
- there were so many pigeons that they were like a huge grey cloud. **(1 mark)**

19 They/Owls fly at night; They/Owls stare; They/Owls look out for pigeons.

(1 mark: two correct for 1 mark)

20 Any plausible, text-based answers are acceptable, e.g. They felt (very) happy/delighted/overjoyed, because:

- the owls had flown off/left.
- they had won the contest.
- there were more pigeons than owls.
- they had shown/proved there are more pigeons than owls. **(1 mark)**

Set A English reading

Paper 2: reading booklet

Why Spiders have Thin Legs

1 Any justified reason, derived from the text, e.g.

   - Anansi was greedy.
   - Anansi loved food/delicious food.   **(1 mark)**

2 black and hairy ☐

   short and thick ☑

   long and thin ☐

   tired and weak ☐   **(1 mark)**

3 Any one of the following:

   - he was very busy/lazy
   - he didn't want to do any jobs/work/sweep the floor/ wash the dishes.   **(1 mark)**

4 beans and rice (Accept a delicious meal)   **(1 mark)**

5 So Monkey could tug when the banana bread was ready.   **(1 mark)**

6 a)

Elephant	— nut roast
Squirrel	— chicken curry
Giraffe	— rice pudding
Hog	— honey cake

   (Elephant → rice pudding; Squirrel → honey cake; Giraffe → nut roast; Hog → chicken curry)   **(1 mark)**

   b) To tug on the web when the banana bread was ready.   **(1 mark)**

7 His mouth watered. (Accept Anansi's mouth watered.)   **(1 mark)**

8 The river washed the webs away. ☐4

   Rabbit tugged a web. ☐2

   Anansi talked to Fox. ☐1

   Anansi's legs started to grow long and skinny. ☐3

   (1 mark: award 1 mark for correct order)

Water

9 washing our clothes. ☑

   burning wood. ☐

   plants to grow. ☑

   drying our clothes. ☐

   (1 mark: award 1 mark for two correct answers ticked)

10 droplets   **(1 mark)**

11 a circle of events that happen again and again in the same order.   **(1 mark)**

12

   (1 mark: award 1 mark for both correct labels)

13 Any one of the following: water droplets move into the air; water changes from a liquid into a gas.   **(1 mark)**

14 blood 80%; skin 70%   **(1 mark: both correct for 1 mark)**

15 Any three of the following: storms; floods; animal waste; food waste; chemicals such as oil; refuse   **(1 mark: three correct answers for 1 mark)**

16 Any two of the following: waste; refuse; litter   **(1 mark: two correct answers for 1 mark)**

17 Any one of the following: only 1% of the world's water is fresh and safe to drink; most of the world's water is salty and we can't drink it; we need water every day.   **(1 mark)**

18

The information says that...	True	False
humans cause all polluted water.		✓
we should try to save water.	✓	
90% of the world's water is not fresh.		✓
you should pick up your litter, but only when you are at a beach, lake or river.		✓
a watering can uses less water than a hosepipe.	✓	

   (2 marks for four correct answers, 1 mark for two or three correct answers)

Set A English grammar, punctuation and spelling

Paper 1: spelling

These are the correct spellings:

1 skate

2 claws

3 trunk

4 nearest

5 feather

6 telephone

7 match

8 pushes

9 Wednesday

10 sandcastle

11 nothing
12 parcels
13 wrist
14 purple
15 quarrel
16 chalk
17 bony
18 useful
19 festival
20 invitation

(20 marks: 1 mark for each correct spelling)

Set A English grammar, punctuation and spelling

Paper 2: questions

a Any suitable present tense verb ending in -ing, e.g. reading, holding, carrying, buying.

b full stop ☐
question mark ✓
exclamation mark ☐
comma ☐

1 (it) is a hot day. (sally) and (tom) are eating ice cream.
(1 mark: all three correctly circled for 1 mark)

2 and ☐   or ✓
but ☐   so ☐   **(1 mark)**

3 on holiday this year! ☐
on holiday this year. ☐
on holiday this year ☐
on holiday this year? ✓   **(1 mark)**

4

Sentence	Full stop	Question mark
You can have tea at our house	✓	
Are you coming to play football		✓
Where are my shoes		✓
It is lunchtime now	✓	

(1 mark: all correct for 1 mark)

5 a verb ☐   a noun ✓
an adjective ☐   an adverb ☐   **(1 mark)**

6 What a wonderful surprise!   **(1 mark)**

7 questions ☐   commands ☐
exclamations ☐   statements ✓   **(1 mark)**

8 Green (plants) need (air), (water) and (sunlight) to grow.
(1 mark: all four correct for 1 mark)

9 happy ☐   inhappy ☐
unhappy ✓   imhappy ☐   **(1 mark)**

10 In the garden, a <u>cheeky</u> robin hopped onto a branch and he sang a <u>sweet</u> song.
(1 mark: both adjectives underlined for 1 mark)

11 Did you see that colour**ful** rainbow in the sky?
The fear**less** explorer went bravely into the jungle.
(1 mark: two correct suffixes added for 1 mark)

12 an adverb ✓   a noun ☐
an adjective ☐   a verb ☐   **(1 mark)**

13 You must bring a coat, gloves, a hat, boots and a packed lunch to school on Friday.
(1 mark: three correctly placed commas for 1 mark)

14 ran (Accept was running)   **(1 mark)**

15 Any response that explains that the apostrophe shows belonging or possession, e.g.
- it shows belonging/possession
- it shows that a/the desk belongs to a/the teacher
- because a/the desk belongs to a/the teacher.

Responses that meet the criteria but use different phrasing should also be marked as correct.

Do not accept responses referring to other uses of the apostrophe, e.g.
- because a letter/letters have been missed out
- because it is short for is/it is.   **(1 mark)**

16 were   **(1 mark)**

17 Award 2 marks for any command sentence that starts with a capital letter and ends with an appropriate punctuation mark, e.g.
- No, don't do that, Brandon!
- That's a bad idea!
- Stop it, Brandon.

Award 1 mark for any command sentence that does not start with a capital letter and/or that does not end with a full stop or an exclamation mark, e.g.
- no don't do that brandon
- you must not do that

Responses that meet the criteria but use different phrasing should be marked as correct.

Correct spelling is not required for the award of the mark. Although correct sentence punctuation is required for the award of both marks, pupils are not required to use capital letters correctly for Brandon or correct internal punctuation.   **(2 marks)**

18 used (Accept was using, had used or has used)

chose (Accept had chosen or have chosen)

(2 marks: I mark for each correct answer)

Set B English reading

Paper I: reading prompt and answer booklet

a in the south-east of Europe

b Any two from: a mainland; mountains; small islands.

I The Severn ☐

The United Kingdom ☑

Europe ☐

The River Severn ☐ **(I mark)**

2 England **(I mark)**

3 Greece has more people than Wales. ☑

Wales has more islands than Greece. ☐

Greece's capital city is Athens. ☑

Mount Olympus is the highest mountain in Wales. ☐

(I mark: award I mark for two correct answers ticked)

4 Any reference to usually hot weather in summer, e.g.

- the weather is usually very hot in the summer

- because it's usually hot in summer. **(I mark)**

5 people ☐   population ☑

capital ☐   million ☐ **(I mark)**

Special Shoes

c in a small village

d

	Lena liked	Maya liked
rounders		✓
hopscotch		✓
reading	✓	
drawing	✓	
dancing		✓

6 because Lena felt lazy ☐

because Lena felt tired ☐

because Lena thought Maya was lazy ☑

because Lena thought Maya was tired ☐

(I mark)

7 Any **two** of the following:

- (because) she could pick enough (by herself)

- (because her) Mum would be pleased

- (so that) they could all eat them with cream for tea.

(I mark)

8 nowhere **(I mark)**

9 "It's your lucky day!" ☐

"Aha!" ☐

"That's for me to know, and for you to find out." ☐

"Hello Lena." ☑

(I mark)

10 curious ☑   frightened ☐

annoyed ☐   hungry ☐ **(I mark)**

11 special shoes **(I mark)**

12 She gasped. **(I mark)**

13 Any answer that quotes or paraphrases the following:

- the soles were clean and shiny, e.g.

   o the soles were new

   o the soles were not worn. **(I mark)**

14 She/the old lady vanished. (Accept she/the old lady disappeared) **(I mark)**

15 warm ☐   very cold ☑

hot ☐   very hot ☐ **(I mark)**

16 tired ☐   surprised ☑

bored ☐   upset ☐ **(I mark)**

17 Any one of the following:

- she poked Maya/her

- she called, "Time to go home!"

- she told her it was time to go home

- she said, "Wakey-wakey, Lazybones!"

- she called her Lazybones and told her to wake up.

(I mark)

18 Any answer that quotes or paraphrases the following:

- because Maya/she blinked

- because Maya's/her mouth fell open.

(I mark)

19 Any one of the following points:

- what else they/the shoes could do

- what/other things they/the shoes could do **(I mark)**

20 Lena put the shoes on. `4`

Lena met an old lady. `2`

Lena's sister fell asleep. `1`

Lena traded the raspberries for a shoe box. `3`

(I mark: award I mark for correct order)

Set B English reading

Paper 2: reading booklet

All about Bees

1 three (Accept 3) **(1 mark)**

2
always female	✓
live for up to one year	☐
mate with the queen bee	☐
die after stinging	✓

(1 mark: award 1 mark for two correct answers ticked)

3 To show the other bees where the food is. (Accept to show other bees where there are flowers with a lot of nectar and pollen) **(1 mark)**

4 Any two of the following: use of chemicals; loss of habitat; lack of food; poisons in the air **(1 mark)**

5 Any two of the following: plants would not be pollinated; some plants would not grow new seeds; some/flower/fruit/nut/vegetable plants would die **(1 mark)**

6 essential **(1 mark)**

7
Pollen helps other flowers to make seeds.	4
The bee moves to other flowers.	2
The pollen rubs off the bee's body.	3
Pollen from a flower sticks to the bee's body.	1

(1 mark: award 1 mark for correct order)

8
Words — **Meanings**

protect — making flowers
habitats — natural homes
flowering — keep safe

(protect → keep safe; habitats → natural homes; flowering → making flowers)

(1 mark)

Alien on Board

9 Any one of the following: their/a teacher; a computer; not really a person (Do not accept very lifelike, strict) **(1 mark)**

10 Any one of the following: it was very quiet; it was very dark; it was very dark, except for the stars; they had never seen any aliens/monsters/creatures. **(1 mark)**

11 They heard a buzz and a loud crash. (Accept they heard a loud noise/loud noises) **(1 mark)**

12
the most furry	☐
the most odd	✓
the most strong	☐
the most friendly	☐

(1 mark)

13 Zoe/she did not like it. (Accept Zoe/she was frightened of it. Do not accept Zoe/she wanted it to go away.) **(1 mark)**

14 terrified **(1 mark)**

15 Any two of the following: she went up to it; she knelt down; she spoke to it gently; she told it not to be afraid/she said, "Don't be afraid"; she said "We won't hurt you!" **(1 mark: two correct answers for 1 mark)**

16 Any reference to the alien belonging to Mr and Mrs Smik, e.g.
- it/the alien belonged to Mr and Mrs Smik
- it/the alien was the property of Mr and Mrs Smik
- its/the alien's owners were Mr and Mrs Smik.

Also accept:
- they need to take it/the alien home to its owners
- it/the alien does not belong to Skylar. **(1 mark)**

17 a)
The story says that Mr and Mrs Smik...	True	False
have five long sticks.		✓
have tentacles.		✓
have very little fur.	✓	
have pale pink skin.	✓	
have lumpy skin.		✓

(2 marks: two or three correct for 1 mark, four correct for 2 marks)

b) Any reference to human beings, e.g.
- they were human beings
- they were humans
- they looked like humans
- they were probably human beings. **(1 mark)**

18 Any reference to dog-dog winking at Skylar, e.g.
- because she is sure dog-dog winked at her
- because she thinks dog-dog winked at her
- because she thinks dog-dog will be back to eat more fungus pie. **(1 mark)**

Set B English grammar, punctuation and spelling

Paper 1: spelling

These are the correct spellings:

1 skill
2 parcel
3 hatch
4 puppy
5 circus
6 clear
7 watches

8 chopping
9 Thursday
10 shy
11 easier
12 health
13 above
14 cabbage
15 hurries
16 wasn't
17 nephew
18 valley
19 ladybird
20 action

(20 marks: 1 mark for each correct spelling)

Set B English grammar, punctuation and spelling

Paper 2: questions

a smallest ☐   small ☐
  most small ☐   smaller ☑

b don't

1 Do you like <u>my new green hat?</u>

(1 mark: all words correctly underlined for 1 mark)

2 My birthday is in (april) and (jane's) birthday is in (may).

(1 mark: three capital letters identified for 1 mark)

3 full stop ☐
  exclamation mark ☐
  question mark ☑
  comma ☐   **(1 mark)**

4 Our school trip is to Leeds Castle.   **(1 mark)**

5 a noun ☐   an adjective ☐
  an adverb ☐   a verb ☑   **(1 mark)**

6 What a great idea ☐
  What a great idea! ☑
  What a great idea? ☐
  What a great idea. ☐   **(1 mark)**

7 We all cheered (loudly) when my sister won the race.
    (1 mark)

8 stayed ☑   stays ☐
  staying ☐   to stay ☐   **(1 mark)**

9

	Present tense	Past tense
I was talking to Mary.		✓
Jack is brushing his teeth.	✓	

(1 mark: both ticked correctly for 1 mark)

10 statements ☐   exclamations ☑
   commands ☐   questions ☐   **(1 mark)**

11 How exciting! We're going on holiday! ☐
   Have you got the tickets? ☐
   Carry your suitcase to the car. ☐
   I have locked the front door. ☑
    (1 mark)

12 Spring, summer, autumn and winter are the four seasons of the year.

(1 mark: both commas correctly placed for 1 mark)

13 We are going to the shop and buying some ice cream. ☐
   We went to the shop and bought some ice cream. ☑
   We go to the shop and bought some ice cream. ☐
   We went to the shop and buy some ice cream. ☐
    (1 mark)

14 When we woke up the next morning, the (soft) (white) snow was falling silently from the sky.

(1 mark: both adjectives circled for 1 mark)

15 He can't go swimming today.
   I hope <u>he'll</u> be better tomorrow.

(2 marks: 1 mark for each correct sentence)

16 They is closing the paddling pool. ☐
   The park is closing soon. ☑
   Peter is closing the door quietly. ☑
   I are closing my eyes and counting to ten. ☐

(1 mark: two ticked correctly for 1 mark)

17

When we got home it was time for bed.		question
Pass the salt.		exclamation
When is it lunchtime?		statement
Thank you! I love it!		command

(1 mark)

18 was trying (Accept tried; had tried)
   went (Accept was going; had gone)

(2 marks: 1 mark for each correct answer)

Spelling test administration

Ensure that your child has a pen or pencil and a rubber to complete the paper.

Your child is not allowed to use a dictionary or electronic spell checker.

Ask your child to look at the practice spelling question. Do this question together.

For each question, read out the word that your child will need to spell correctly.

Then read the whole sentence.

Then read the word again.

Your child needs to write the word into the blank space in the sentence.

Here is the practice question:

The word is **play**.

The children **play** in the park.

The word is **play**.

Check that your child understands that 'play' should be written in the first blank space.

Explain that you are going to read 20 sentences. Each sentence has a word missing, just like the practice question.

Read questions 1 to 20 to your child, starting with the question number, reading out the word followed by the sentence, and then the word again.

Leave enough time (at least 12 seconds) between questions for your child to attempt the spelling. Do not rush, as the test time of 15 minutes is approximate and children will be given more time if they need it.

Tell your child that they may cross out the word and write it again if they think they have made a mistake.

You may repeat the target word if needed.

Set A English grammar, punctuation and spelling

Paper 1: spelling

Practice question: The word is **play**. The children **play** in the park. The word is **play**.

Spelling 1: The word is **skate**. The ice is too thin to **skate** on. The word is **skate**.

Spelling 2: The word is **claws**. Our cat has long **claws**. The word is **claws**.

Spelling 3: The word is **trunk**. An elephant's nose is called a **trunk**. The word is **trunk**.

Spelling 4: The word is **nearest**. Sit in the chair **nearest** the door. The word is **nearest**.

Spelling 5: The word is **feather**. Sally found a magpie **feather**. The word is **feather**.

Spelling 6: The word is **telephone**. Please give me your **telephone** number. The word is **telephone**.

Spelling 7: The word is **match**. We all enjoyed the cricket **match**. The word is **match**.

Spelling 8: The word is **pushes**. Dad **pushes** Milly's pram. The word is **pushes**.

Spelling 9: The word is **Wednesday**. We are going swimming on **Wednesday**. The word is **Wednesday**.

Spelling 10: The word is **sandcastle**. We build a **sandcastle** on the beach. The word is **sandcastle**.

Spelling 11: The word is **nothing**. Hannah has **nothing** to do today. The word is **nothing**.

Spelling 12: The word is **parcels**. Jack takes the **parcels** to the Post Office. The word is **parcels**.

Spelling 13: The word is **wrist**. You wear a watch on your **wrist**. The word is **wrist**.

Spelling 14: The word is **purple**. The hat is **purple** with yellow flowers. The word is **purple**.

Spelling 15: The word is **quarrel**. I do not want to **quarrel** with you. The word is **quarrel**.

Spelling 16: The word is **chalk**. Please pass me the pink **chalk**. The word is **chalk**.

Spelling 17: The word is **bony**. This fish is very **bony**. The word is **bony**.

Spelling 18: The word is **useful**. Plastic is a **useful** material. The word is **useful**.

Spelling 19: The word is **festival**. Our school has a **festival** in September. The word is **festival**.

Spelling 20: The word is **invitation**. Alex gave me an **invitation** to his party. The word is **invitation**.

Set B English grammar, punctuation and spelling

Paper 1: spelling

Practice question: The word is **flew**.
The birds **flew** into the sky.
The word is **flew**.

Spelling 1: The word is **skill**.
Gina shows great **skill** on the piano.
The word is **skill**.

Spelling 2: The word is **parcel**.
The postman brought a big **parcel**.
The word is **parcel**.

Spelling 3: The word is **hatch**.
The eggs **hatch** under the hen.
The word is **hatch**.

Spelling 4: The word is **puppy**.
Our new **puppy** loves to play
with us.
The word is **puppy**.

Spelling 5: The word is **circus**.
The children had an amazing time
at the **circus**.
The word is **circus**.

Spelling 6: The word is **clear**.
The sky was **clear** when we left
home.
The word is **clear**.

Spelling 7: The word is **watches**.
Tom **watches** TV after he has done
his homework.
The word is **watches**.

Spelling 8: The word is **chopping**.
Dad is **chopping** vegetables for
dinner.
The word is **chopping**.

Spelling 9: The word is **Thursday**.
Gran is coming to tea on **Thursday**.
The word is **Thursday**.

Spelling 10: The word is **shy**.
Jane was too **shy** to join in our
game.
The word is **shy**.

Spelling 11: The word is **easier**.
He thinks reading is **easier** than
writing.
The word is **easier**.

Spelling 12: The word is **health**.
Eating fresh fruit every day is good
for your **health**.
The word is **health**.

Spelling 13: The word is **above**.
The hot air balloon rose **above**
the rooftops.
The word is **above**.

Spelling 14: The word is **cabbage**.
She loves all vegetables, especially
cabbage.
The word is **cabbage**.

Spelling 15: The word is **hurries**.
Mum **hurries** to answer the door.
The word is **hurries**.

Spelling 16: The word is **wasn't**.
Jane **wasn't** at school today.
The word is **wasn't**.

Spelling 17: The word is **nephew**.
Uncle Toby took his **nephew** to
the park.
The word is **nephew**.

Spelling 18: The word is **valley**.
The river runs through a deep **valley**.
The word is **valley**.

Spelling 19: The word is **ladybird**.
Yesterday I saw a **ladybird** with seven
spots.
The word is **ladybird**.

Spelling 20: The word is **action**.
I enjoy reading stories with lots of
action.
The word is **action**.